The Tin Man

The Voice of an Incest Survivor

Written by Holly Otten

Photographs by Steven Otten and Holly Otten

ISBN 978-0-615-29397-4

Text, art/illustrations, photographs, and book design by Holly Otten, unless otherwise noted.

Printed in the United States.

"Open Heaven" Acrylic by Holly Otten p. 15

"Rose" oil by Holly Otten p. 21

"For My Boys" acrylic by Holly Otten p. 24

"Real Joy" cartooning by Holly Otten p. 38

"The Tin Man" pen and ink by Michael Forcino p. 23

"A Second Chance" poem by Michael Forcino p. 46

Cover: Art by Stephen Forcino

Email: tinmanministries@cox.net

I dedicate this book

To the Master Healer;

Giver of Life; God of Love...

To the reader who has been a victim of abuse—

Be a victorious survivor...

Lastly, but not in the least,

To my husband Steve, and "my little men"

Michael, Stephen, and JP

Acknowledgments

Much thanks to my family…

Much thanks to my oldest son, Michael, for sharing his God-given talent for art. Michael also contributed a poem which he proudly displayed to me following school one day, just days before printing. It is wonderfully profound, and depicts the spirit of The Tin Man. I feel that it is an appropriate and complimentary addition to this book.

Thanks to my son Stephen, for generously sharing his artistic talent. I love his interpretation of The Tin Man on the cover.

Thanks to my youngest, JP, for sharing his angelic and curious face for two of my photo shots. The sunflower he sniffed was two times his size.

And lastly, I am grateful to my husband Steve, for sharing his love of nature which is so remarkably displayed in his photos. He has an amazing gift to catch the wonders in nature, from its smallest, simplest traits, to its largest and most grandeur. Because of Steve, I am beginning to "see" the world with new eyes. My life has been enriched by the moments when I am able to commune with nature and experience its wonders.

Prologue

The Tin Man in the classic movie "The Wizard of Oz" traveled high and low, through valleys, and treacherous forests. Why? He wanted a new heart. He was told that the great Wizard of Oz could give him a new heart.

I spent the first half of my life with a broken, wounded heart. I was full of fear, anger, disappointment, shame, and mistrust. My heart was not filled with love, peace, and happiness. I wanted to be happy so badly. I wanted to be loved.

Someone told me when I was twenty-three years old that Jesus loved me. I ventured on the journey toward the great and most powerful Wizard of all, Jesus Christ, to ask Him to give me a new heart. After all, I had nothing to lose. Nothing else had made me happy.

That journey began with the uncovering of my wounded heart. It had to be given a chance to heal. My journey was much like that of The Tin Man. There was much turmoil, and many trials that I had to face. These pages bear the raw emotions that I experienced while receiving counseling. They came straight from the journal I kept during this period. I found that it was all worth the effort. Eventually, my heart became healed. It became whole. It became new. Bear with me as I lead you through my journey. Just remember there's a happy ending, just like in the movie.

Part One

The Journey Begins

"This is going to hurt a little,"
says the mommy to her little girl
as she tears the band-aid off her
infected wound.

TELL IT...

In an answering machine message...

Hello, you've just reached Holly, an incest victim. I can't talk to you right now. I'm too busy baptizing the floor with my tears. Even if I weren't crying, I'm much too miserable to talk to you. I'd be bound to make you feel the same. And I really don't feel like making you happy.

So leave a message at the tone, preferably an "I love you," or an "It'll be OK." I'll get back to you when I'm dried out and jovial.

In a speech...

Hello. My name is Holly. I am an incest survivor. I was molested by my father, and my brother. My great-grandfather too. When I was a teen, a near-stranger had his way with me. I've spent most of my life trying to prove I'm somebody with feelings; a somebody like everybody who craves to be loved and to love in return. Even my mother didn't show me love. Not only was she never home to show me her love, but she just didn't know how to love me.

Today, I am twenty-six years old and just learning to love and be loved.

Do You Have the Answers

I'm drowning
I'm trying to stay alive
Where is the energy
Booze, drugs, sleep, God

I'm tired
Of pain
I want to be happy
Does anyone really know love
If they do, are they really happy

Confusion

There's another world
Inside my heart
It's a lonely world
But oh so full
Emotions rage and run rampart
They run deep—to the core.

Someone just stabbed me
In the pit of my stomach
Those damn emotions
They get in the way
Are they good or bad
Who is good
Who is bad
Will someone tell me
A child does not know

It's dark and confusion settles in
My head is dizzy
I'm getting weak
Will someone help me
I'm getting sick
Help me

Words of Comfort

"Thus says the Lord,
Who created you, and
Formed you,
'Fear not,
For I have redeemed you.
I have called you by name.
You are mine...
You are precious in my
Eyes, and
Glorious,
I love you." Isaiah 43:1, 4

"Thus they shall know that I am the Lord when I break the bonds of Their yoke and free them from the Power of those who enslaved Them...they shall dwell secure, With no one to frighten them."
Ezekiel 34:27

"With all my ways you are Familiar. Even before a word is on My tongue, Lord, you know it all.

Behind and before me you Encircle me, and rest your hand Upon me."

Psalm 139:4, 5

"Who will separate us from the Love of Christ? Will distress, or Tribulation, persecution, or Nakedness, peril, or sword? But In all these things we Overwhelmingly conquer through Him who loves us."
Romans 8:35, 37

Part Two

Purification

"This will help
your boo-boo get better,"
the mommy says
as she applies the disinfectant
to her little boy's wound.
"Ouch!" he replies.

A Cut Flower

by Karl Shapiro (1913)

I stand on slenderness all fresh and fair,
I feel root-firmness
In the earth far down,
I catch in the wind and
Loose my scent for bees
That sack my throat for kisses
And suck love.
What is the wind
That brings thy body over?
Wind, I am beautiful and sick.
I long for rain that strikes
And bites like cold and hurts.
Be angry, rain, for dew is kind to me
When I am cool from sleep
And take my bath.

Who softens the sweet earth
About my feet,
Touches my face so often
And brings water?
Where does she go,
Taller than any sunflower
Over the grass like birds?
Has she a root?
These are great animals
That kneel to us,
Sent by the sun perhaps
To help us Grow.
I have seen death.
The colors went away,
The petals grasped at nothing
And curled tight.
Then the whole head fell off
And left the sky.

She tended me
and held me by my stalk.
Yesterday I was well,
And then the gleam,
The thing sharper than frost
Cut me in half.
I fainted and was lifted high.
I feel waist-deep in rain.
My face is dry and drawn.
My beauty leaks into the glass like rain.
When first I opened to the sun
I thought my colors would be parched.

Where are my bees?
Must I die now?
Is this a part of life?

LISTEN

Don't touch me
I'm alive!
Hey you monster
I know I'm real small
I can't talk
I can't tell you, "STOP!"
I can't tell on you either
But I can feel!
I think you've confused me with a doll
They are not alive
They can't feel
They can't hurt
They don't have opinions
But I do!

So listen up!
I'm a person just like you
I have a heart and a mind
And if you pricked me with a pin
I'd feel it.

And if you touch me
My mind I will lose
My heart it will break
I will die

Don't touch me, I said
"I'M ALIVE!"

WHO AM I?

Escape the pain
Escape
Any way you can

Be a good girl
Be a bad girl
Be a party girl

You are a fool
There's no escaping it...

THE PAIN!

I Will Survive

I think of the rose
How in the winter
It appears as if it's dead
Everything turns brown
The leaves dry up
And fall to the ground
You can cut its branches
All the way down
But it's just asleep for a night
In its bed
It's alive
It survives

In the morning when the sun
Sheds its warmth
And touches the roots
Down below
When the water from the snows
Gives new life through its toes
A new rose is born
The rose awakens with new life
It still has life to give
For it lives to give
It offers me hope
I will imitate the rose

Words of Hope

"Because he clings to me, I will deliver him; I will set him on high because he acknowledges my name. He shall call upon Me, and I will answer him. I will be with him in distress. I will deliver him and glorify him; with length of days I will gratify him, and will show him my salvation."

Psalm 91:14-16

"Though my father and mother forsake me, yet will the Lord receive me."

Psalm 27:10

"I have stilled and quieted my soul like a weaned child. Like a weaned child on its mother's lap, so is my soul within me. Oh Israel, hope in the Lord, both now and forever."

Psalm 131:2

"For I hope in you O Lord, you will answer O Lord, my God."

Psalm 38:14-16

"Be strong and let your heart take courage, all you who hope in the Lord."

Psalm 31:24

"For I know the plans that I have for you; plans for welfare and not calamity; to give you a future and a hope."

Jeremiah 29:10

THE CHAIN

MY LIFE HAS BEEN STOLEN
SNATCHED AWAY
I WAS NEVER A CHILD
WHAT IS IT LIKE
I WANT TO BE ONE

SOMEONE DIDN'T LIKE THEIR LIFE
SO THEY STOLE MINE
HOW SELFISH MAN CAN BE
MEN—THEY CAN BE REAL SELFISH
THEY CAN BE REAL STRONG
BUT NOT INSIDE

THAT'S WHY THEY'VE TAKEN MY LIFE
TO MAKE THEMSELVES STRONG
THEY ARE WEAK
I AM STRONG!
I AM GOING TO WIN MY LIFE BACK
WHILE THEY ARE STILL WEAK

IT'S A CHAIN
BUT NOT ANYMORE
THIS CHAIN WAS MADE TO BE BROKEN
IN BITS AND BITS
THROWN DOWN INTO HELL
WHERE IT BELONGS
WHERE IT CAME FROM

Milton lost his paradise
Dorothy lost her way
Vincent lost his sanity
Thomas lost his faith

Hoover lost the second time
Sigmund lost his friend
Me, I lost my innocence
And I want it back again

Lyrics sung by Susan Ashton
"Angels of Mercy"

Princes and Princesses

There was once a town that was quite shabby. The town was a reflection of the people who lived therein. One day, a king visited the town and brought his new born baby. While in the town, he switched his baby with another. Then he left the town. The people did not know which baby was the king's. They were so afraid to treat the king's child poorly. They were afraid that upon the king's return, he would be angered toward them.

Consequently, the entire town treated all of the children with royalty.

Within a few years, the town had changed considerably. It was with libraries, schools, and hospitals which helped to nurture and care for the children. Within a few more years, the children grew up. They were healthy, loving and inspiring adults.

Finally, the king returned. And when he did, he visited a ninety year old woman on her death bed. The woman had to ask the king, "Is my daughter the princess?" He replied, "*All* of the children are princes and princesses."

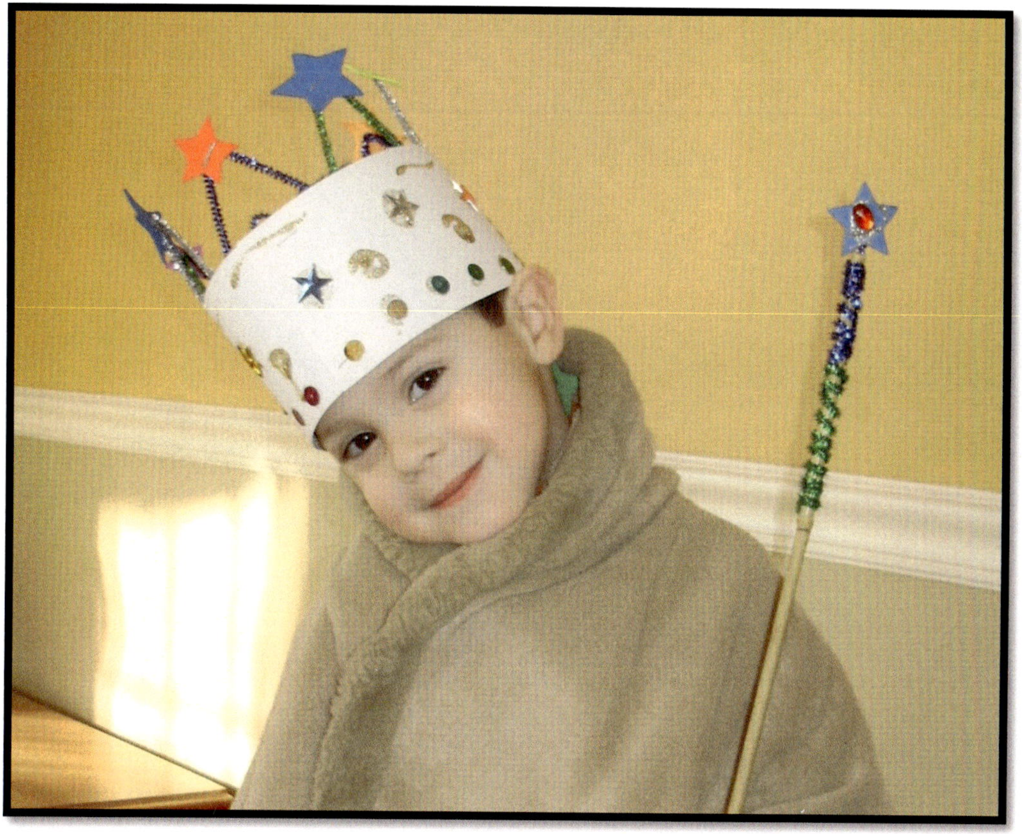

"See that you do not despise one of these little ones, for I say to you, that their angels continually see the face of My Father who is in heaven."
Matthew 18:9-11

"We know that our old self was crucified with Him...
Christ was raised from the dead by the glory of the
Father, so that we too might live in newness of life."
Romans 6:4, 6

Like Water

When water is trapped
It cuts itself a new path
It finds a new way
Nothing can keep it from flowing
It cuts its way through
Rock, wood, and iron
It is strong
It perseveres
It is powerful

I want to be like water
When it seems as if
I can't go any further
When I feel as if
There's no way out
When it feels like
The end is at hand
I will cut me a new path
Like water

IF YOU WANT…	**YOU MUST…**
PEACE…	*FORGIVE*
JOY…	*FORGIVE & ACCEPT*
LOVE…	*FORGIVE & LOVE YOURSELF*
HAPPINESS…	*FORGIVE & LET GO OF BITTERNESS*
REST…	*FORGIVE & SIT WITH YOURSELF*
FAITH…	*FORGIVE & BELIEVE*
STRENGTH…	*FORGIVE & DEPEND ON GOD*
TO TRUST…	*FORGIVE & JUST DO IT*
HEALTHY RELATIONSHIPS…	*FORGIVE & HEAL*

Beyond Justice to Mercy

"I don't know if now is the time
To surrender the silence
Between your heart and mine
But the love that I've chosen
Cries out to be spoken
Leaving the heartache behind
We must reach out beyond
Justice to mercy
Going more than halfway to forgive
A tender voice is calling me
To that place of compassion
Where hearts run pure and free
Where the hunger for vengeance
Gives way to repentance
Where love will teach us to see
We can reach out beyond
Justice to mercy"

Lyrics sung by Susan Ashton
"Wakened by the Wind"

Part Three

It is Healed

The wound is gone. The pain is gone.
The little boy jumps up and down with glee.
"Thank you Mommy for making me all better."

Dear Adult,

I realize that I've done things in which you don't approve of. But you see, I was made dirty from the beginning. It made it easier to do the things in which I chose to do. It was also my way of getting attention. So there was an innocence; a longing to be loved. You want that, don't you? But you try to get it by being good. I do want to be good. And I want to be made clean. I want to start new. Will you help me? Will you forgive me for disappointing you?

Love,

Inner Child

Dear Inner Child,

I am sorry for hurting you, and for not understanding. You've been hurt enough. You do not need me condemning you. You just need love, and to be healed of all the wounds in which you have been afflicted. You're not truly mean inside. You do have the ability to love. And there's nothing to be afraid of. I will be there for you. I will try to love you and care for you. I forgive you Inner Child. How can I condemn you when our Lord doesn't condemn you?

Love,

Adult

My Daughter,

I am Jesus, your Savior. I am the Lord, your God. I am your Creator. I created you, and so you are beautiful. My Son came and gave His life for you, and you therefore, are my daughter. The things that you have done in your past were not so pleasing to me, but you have repented of your sins, and have come to Me. Even when your behavior wasn't pleasing to Me, I still loved you and you were Mine.

My Daughter,

I have forgotten the past. My Son's death has given you life. Start rejoicing and living this new life. You are so pleasing in my sight. Rejoice! I have made you whole new. You are a new creation. Thank you for accepting my invitation. I am glad you came back to Me.

I love you my precious child. Never leave my side. I will protect you and guard you.

Happy Birthday!

Skim the Fat

"Why do you want to bring up the past?" I've heard people say after I told them I was seeing a counselor. I started seeing a counselor eighteen years ago when my memories of incest came to the surface. I wasn't seeing a counselor prior to that moment. I didn't go looking for "trouble" or a reason to wallow in sorrow. I actually felt that I couldn't handle the stress and the pressure all on my own. When I found a counselor to confide in, I didn't feel alone. She provided not only an ear to listen, and a shoulder to cry on, but she had the knowledge and skills in how to respond to me and my emotional needs. She knew when to *just* listen. She knew what words I needed to hear. She had lots of wisdom that I lacked.

These abilities she possessed were assisted and fueled by the Holy Spirit. My counselor was Christian. We prayed together prior to our sessions, and again at the conclusion. We invited the Holy Spirit and thus we welcomed all the graces that He offered, such as His wisdom, knowledge, strength, comfort, light and truth. I can testify to His power in those sessions. Healing passed swiftly and powerfully through me.

Did I welcome the pain that accompanied the healing process? Certainly not! But I didn't run away from it. I took on my past as a fighter does in the boxing arena. I wanted to conquer all that haunted me...my recurring nightmares, countless dead-end relationships...seeing my father in the men I dated...an endless pit of loneliness...never being satisfied or happy. I wanted to be rid of it all—rid of the "fat."

When I was a young girl, I remember the first time I saw my Nana "skim the fat." She took the day-old pot of chicken soup out of the refrigerator. I thought, "Boy, that doesn't look like the chicken soup I saw her make yesterday." She saw the look of question on my face, and then explained to me that she was removing the fat that rose to the top and then hardened. She preferred to throw the unhealthy fat into the garbage rather than eat it. By the looks of it, I had no objections. It didn't look very appetizing to me.

Since that day, I've made chicken soup of my own. I've noticed that if I don't remove the day-old fat from my soup and just reheat it, the layer of fat seems to disappear. But it really doesn't. When the soup is heated, the fat liquefies and blends in with the water.

With this analogy I hope to explain the affects that our childhood wounds have on our lives. Our childhood experiences whether good or bad, taint our view of things. In my case, my deepest wounds came from the abuse of men. My father, brother, great-grandfather, and near-stranger wounded me. Later as an adult, I harbored feelings of hate, anger, and fear towards men. It kept me from having happy, healthy relationships with men. If I continued to keep those wounded feelings squelched inside me, they would still resurface in the way I viewed things and people.

I know this to be true. Before I underwent counseling, I had horrible relationships with men. I could never allow a man to get close to me. I feared they would hurt me, abuse me, abandon me, or see me as I truly was (which wasn't a good opinion). Another part of me hated men and didn't want to be close to one. Those feelings led me to want to abuse and hurt men any way I could.

Through counseling, I became aware of these hidden feelings. I came to accept them and feel them for what they were. Then I became ready to let go of them. I released them. I didn't choose to let those ugly, dark feelings abide with me anymore. I longed for happy feelings. I longed to be filled with love. I prayed to God to fill me with His love. I didn't know what love was or how it was suppose to feel. I didn't know how to be happy. What did it mean to live in peace? I remember my first *real* laugh. I was in my early twenties. I made a sound that erupted from the depths of my soul. The sound was a joyful one. It described the feeling that I had within. I was so surprised to hear myself make that unfamiliar sound. "So this is what it feels like," I thought. "I had only pretended to laugh all these years." In the past, my laughter was superficial.

Because I let go of the ugliness inside (I skimmed the fat and disposed of it) I feel free to be me—free to let you see me—the me that is within. I am not squelching or hiding parts of me. I am now able to be truly me. This fosters a sense of peace within.

I will never agree with the adages, "It's not good to delve into the past," or "Leave the past in the past," as I've heard so many say. If the past keeps resurfacing, then it hasn't been dealt with yet. The past always finds a way into the present and the future if there are unresolved feelings. And *that* is not good. Similarly, if the fat is skimmed from the day-old soup, then there won't be a layer floating on the top of the two-day old soup.

A New Name

NEW. It's a terrific word. It provokes happy feelings and thoughts. New cars. New dress. New iPOD.

OLD. Yuk. Ditch it if it's old. It's worn out. No good. I've done that. I've heard that. Seen that.

Would you trade in your "old" self for a "new" self? Cinderella was given that choice for a night. I feel like I've lived the tale of Cinderella. My wish was made real by a powerful, magical person named Jesus Christ.

My name is Holly, but I feel more like a "Grace." God gave me a new name—a new identity. I like this new person that I've become, ONLY by the grace of God. In other words, as the Encarta Dictionary defines "grace," I received "the infinite love, mercy, favor, and goodwill shown to humankind by God."

When I was younger (my OLD self) I didn't like myself. To begin with, my own parents seemed to not like me. I was abused by my father and neglected by my mother. When I was ten I sought to be liked by my peers. I gave in to peer pressure and began smoking marijuana. By the age of twelve I was taking pills. At fifteen I was snorting cocaine. I followed my peers by skipping school days at a time, hung out at places until 1:00 a.m., and socialized with twenty year old men.

At seventeen I lost my virginity to an older man on our first date. I never saw him again. That was the beginning of a promiscuous lifestyle. I was innocently seeking to be liked by others. I was a tall, thin, intelligent, and attractive young girl. I got lots of attention from men. I reveled in that attention. Deep inside I was longing to be loved. Actually, I was validating the inner knowledge that I was no good, that I was shameful.

At twenty I began stealing money from the companies I worked for. I dated men continuously. Every man that was good to me and who began to show true love for me, I abused. I felt unworthy of kindness so I ended every relationship that I had by responding with unkindness.

At twenty-three I met people who believed that Jesus Christ was their Lord and Savior. They were such happy and kind people. They showed that they cared for me. It was then that I opened my heart for a third time to Jesus. I welcomed Him into my life once again and haven't turned away since. I did that because inside I was pleading for help. I felt lost and lonely. When I invited Jesus into my life, I suddenly felt that I was loved. It was an unconditional love. I became sorry for all the wrong things that I had done. I asked Him to help me to change and become a better person. I never felt that I was unworthy of His love. I never felt judged by Him. I only felt an unconditional love, that which I didn't receive as a child from my own parents.

It was at this time that I reclaimed my virginity. I fell once or twice, but I got up again and pushed forward. I've had to forgive myself many times over. Jesus' love has helped me to love myself. Jesus has restored my innocence; the innocence of my youth. I've grieved the loss of my youth. But I celebrate my new life. I do so many of those childlike things now that I once longed to do. I am happy with the simple things in life, just like a child.

Want a new name? Just ask Jesus. In 2 Corinthians 5:17, the Word tells us, "If anyone is in Christ, he is a new creation. The old has gone. The new has come." What would your new name be? How about Hephzibah? In Isaiah 62:4, God's Word states "You shall no longer be called (or named) Forsaken...but you shall be called Hephzibah (meaning "my delight is in her"). Personally, I prefer Grace.

Hershey Kisses

Is Jesus still alive all these years later? Has anyone seen Him? People talk of miracles. I just heard from a friend that her friend went to her Oncologist and found that her malignant tumor had disappeared. I also heard that a woman who was thought to be barren, and thus adopted two children, just became pregnant. They say that Jesus is the reason. You mean the Jesus who walked on the earth 2000 years ago?

I don't see the wind, but I can see the results of its power as it drags a tree down in my yard. When I was thirteen I witnessed my baby cousin enter the world right before my very eyes. I felt as though I witnessed a miracle. I thought that there had to be a God to have created this wondrous, beautiful gift. So maybe Jesus ***is*** still alive and ***is*** still performing miracles. My experiences have taught me that if one opens his eyes of faith, he will see God's hand at work in his life. Faith—Prayer—move the hand of God.

Repeatedly throughout Jesus' brief period of evangelizing, He worked miracles. Almost every time following a healing, Jesus said, "Your faith has healed you." People in need would approach Jesus. He would ask them what they wanted of Him (even though He already knew). He quite often was moved and filled with empathy. Out of love, He answered their prayers and they were healed.

At the time of my conversion, I prayed to God; not knowing for sure if there really was a God who exists, who knows who I am, and hears my prayers. I was sad and lonely so I cried out to God every night. God heard. He responded to my prayers. "Yes," I thought, "there is a God who loves me".

From that point in my life, God blessed me with the gift of faith. I didn't doubt that He was present in my everyday life. I didn't doubt that if I spoke to Him in my heart that He heard me. For twenty years now, my faith has repeatedly seen God's presence in the simplest moments and in my deepest, darkest moments. These moments in my life can be described as miracles. A supernatural power or *grace* intervenes in my mortal world. I like to think of these "divine" moments as Hershey Kisses, handed to me by God, my loving Father.

Imagine this scenario: There is a father who has a ten year old daughter. Every now and then he finds himself stopping at the local milk store, and decides he'd like to make his daughter smile by giving her a sweet treat. He eyes the candy isle and spots a bag of Hershey Kisses. He gets home from the store and greets his daughter with a hug and hands her a Hershey Kiss. A huge grin explodes on her face. She knows that her father did this as a sign of his love. She is touched and wraps her arms around her father and gives him an even bigger hug than the first one. This is her way of saying that she loves him too.

In the past few months I can think of two of these moments in which God my Father gave me a Hershey Kiss. The first one happened on Easter morning. Prior to that morning, I had been experiencing a very difficult time. Ironically, it was the Church season of Lent. I had been experiencing severe symptoms of Chronic Fatigue and Fibromyalgia, and also depression. I felt as if I were living in a dark and gloomy world. Many times I found myself turning to God for help. I prayed that He would shine His light into my darkness.

During Lent, I had passed a church that displayed a sign by the road that read, "Sunrise Service on Easter morning." I had always wanted to attend a sunrise service on Easter. I had never been to this church before, but on Easter morning I woke my husband and three small children at 6:00 a.m. to attend the sunrise service. The temperature was in the high 30's. I bundled up my children and brought blankets to wrap them up. It was a beautiful, joy-filled service with the Living Passion.

As we were leaving the service, I noticed the sun rising above the tree line. I thought it appropriate to take a picture of the sunrise. I took two pictures. Later that day I downloaded the pictures on my computer. As my husband and I viewed the pictures of the sun blown up on the screen, we noticed something curious about the second picture that I took. There was a white image in the sky. My husband thought that it looked like the Risen Christ. In that same picture, there were sunbeams that formed a cross and a red diamond appeared on the cross. None of these things appeared in the first picture which I took a minute earlier.

I prayed and asked God if this was a sign of His presence and He brought to my mind the story of "The Transfiguration," from the Bible. There is a verse that states, "His (Jesus) face shone like the sun and His clothes became white as light," Matthew 17:2. Immediately I felt that God had given me a Hershey Kiss. I had been praying for His light and He came in the form of light, during the rising of the Easter sun.

Several months later my husband and I celebrated our anniversary by vacationing at an inn on the harbor. While there, I had taken pictures of every sunrise and sunset. I took pictures of everything and everyone it seems. One evening my husband and I decided to walk the beach under the light of the moon. It was the only time that I didn't have my camera. I had reluctantly forgotten it back at the inn. The moon was making a grand show of itself. I so desired to take a picture of it.

The next morning I rose to take pictures of the sunrise. After the sun had fully risen and it shown amidst a bright blue sky, I got up to put my camera away. However, from a different view, I spotted the moon which still shown faintly in the sky. I decided to take a picture of it. I don't know why I did. There was nothing special about it. But it really was special after all. When my digital camera revealed the picture in its small screen, it depicted a bright moon in a black sky. It was hard to believe what my eyes were showing me. I showed my husband and then the caretakers of the inn, both of whom possess knowledge of photography. No one could explain the picture scientifically. I knew then that my Father in Heaven, who is with me on this earth, handed me a Hershey Kiss. God gave me the picture of the moon that I so wanted the night before. Maybe it was His anniversary gift to me and my husband.

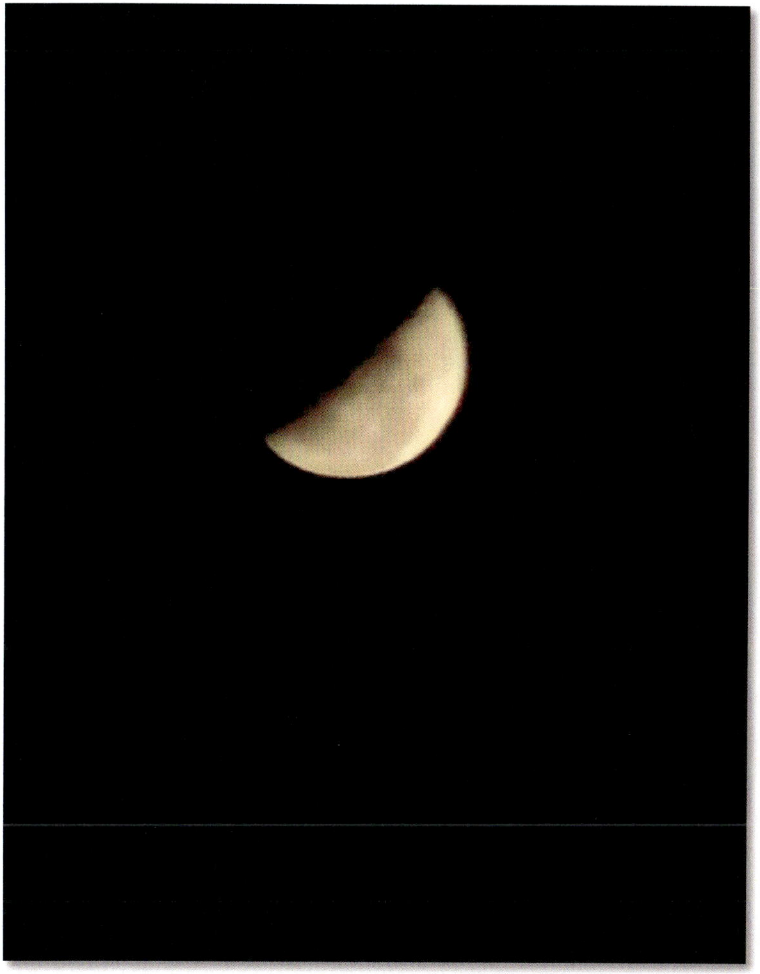

Did I do anything to deserve these "Kisses?" Am I special or something? I do not feel that I am more special than any other human being that God created. We are all His children. He loves us all the same. None of us deserve His attention or love more than another. I think that God has Hershey Kisses for all of His children. Unfortunately there are some who blatantly deny the Father's love, and thus His gifts. Others don't think that they deserve a gift from their Father, so they are not open to receiving it.

Me? I'm a chocoholic. I have my hands out begging my Father for more "Kisses." Life is full of disappointments, so God, who is the same God that performed miracles in the life of Jesus Christ, gives us moments to encourage us. All we have to say is "Thank you God."

After thought: On the day that I sat at my computer writing this piece, I received a gift in the mail. My aunt sent me a plaque, "Just because." It says, "I never met a chocolate I didn't like." Coincidence or God incidence?

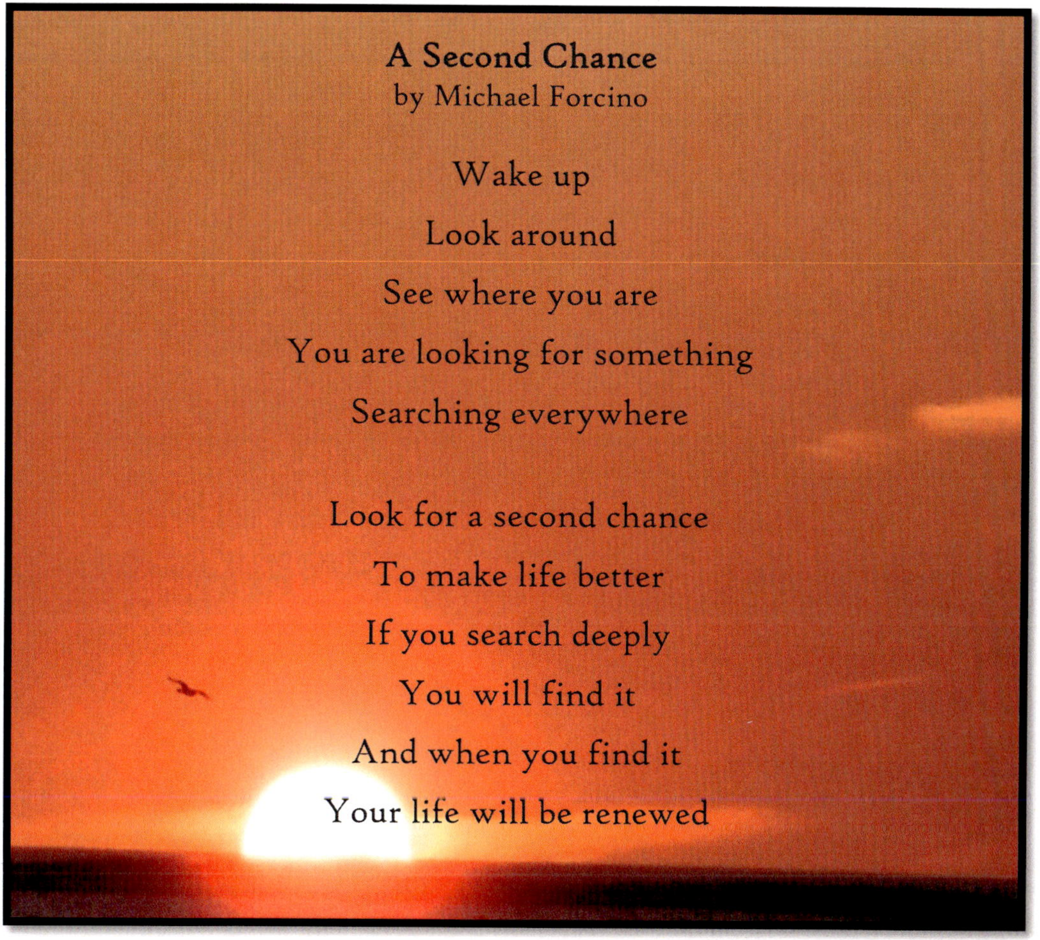

A Second Chance
by Michael Forcino

Wake up
Look around
See where you are
You are looking for something
Searching everywhere

Look for a second chance
To make life better
If you search deeply
You will find it
And when you find it
Your life will be renewed

Well Water

Based on the "Samaritan Woman"
John 4:7-19

There was a girl who grew up without the love of her father and mother. Her father abused her, and her mother neglected her. They were wounded themselves and unable to love the little girl.

The girl grew up. She lived in Samaria. As a grown woman she didn't think very much of herself. In addition to not being nurtured as a child, she was a Samaritan, and a woman. Each distinction was considered inferior in her culture.

The woman wanted so much to belong to someone and to be treated with love that she married. Eventually she married five times. By the time of her sixth relationship, she decided that they would only live together. But even then, she still felt empty and unloved.

But a miracle happened.

One day as the woman approached the well in the middle of town, she met a man. He asked her for a drink of water. Then he told her that he knew where she could find water to drink, which would take away her thirst forever. If she would drink of the water that he offered her, she would have everlasting life.

The woman wanted to excuse herself to go tell her mate back home about this life-giving water. The man told her that he knew she had married five times, and was now living with a man. With such knowledge, the woman knew that he was a prophet. She asked Him for the water that only he could give so that her thirst would be quenched forever. The man was very pleased to oblige her. He knew that the woman would never be the same. She would be fulfilled. Finally.

It happened as the man said. The woman found a contentedness in her life that she had never known. She no longer felt empty and alone. No man was able to satisfy her except for the man at the well.

The man's name was Jesus.

SEX

SSHHH. This is a private matter, one that should be shared in a private place. However, I feel like shouting from a rooftop how blessed I am to finally understand what the hype is all about. I find it difficult to describe how wonderful it is for me. It must be a divine gift from God up above, as I heard from the nuns who used to teach me. It is a shame that only now, at the age of forty-three, have I come to this realization.

SEX. Is it good? Is it bad? It depends on who you ask. If you ask me now, married for the second time, after years of counseling, healing and "deprogramming" by my Lord Jesus, I would say, "It is one of the most magnificent things created on this earth; a glorious gift given by our Creator."

Since I have had this new understanding, I prefer to use the other term that is used for sex, "to make love." You've heard it all before, "You should save yourself so you can share it with your spouse, the one to whom you are committed to for the rest of your life." I have discovered the wisdom in that belief. Just the act itself can be a lot of fun, but what makes it so incredible is when it is shared with your life-long bride or groom.

However, sex is never fun or beautiful if you've been sexually molested or raped. Because I was molested on several occasions, I never liked sex, even though I lived a promiscuous lifestyle. I never initiated it. I went along with my partner so that I would please him. I was getting attention and affection which I so longed for.

Eventually, I started living a new life in Jesus Christ and reclaimed my virginity. For seven years I saved myself for my husband. I thought that sex or "making love" would be everything I heard it could be between a husband and wife. Beginning on my wedding night, I found that it wasn't that way. Over time I came to understand that although I had worked through years of counseling, I hadn't dealt with the issue of intimacy. Intimacy is a huge undertaking for a person who has been sexually abused.

I came to understand that I feared intimacy. I realized a pattern in my behavior with my husband before going to bed at night. I very frequently started arguments at bed time. I realized that I did this as a way to keep my husband from getting intimate with me. I felt more comfortable facing away from my husband, while laying on the farthest edge of the bed. When we were intimate, I discovered how very tense I was. I couldn't relax and enjoy myself. Sadly, it wasn't fun, and it wasn't beautiful.

God very gently enlightened me and graciously helped me to heal. I no longer fear intimacy. I embrace it. I foster it in my marriage. I revel in it. I realize now why sex is also known as "making love." When a person loves someone as deeply as one loves a spouse, it is difficult to express that love in words. When those two people share the intimate moment of "making love," those feelings of love are expressed in the utmost.

The intimacy that I now share with my husband is one of the greatest gifts that God has ever given to me. This gift is for all, including those who have been abused and those who have regretfully abused the gift.

"Sinner." I am afraid to say the word around others. It seems as if it is taboo. Some people say, "I'm not a sinner. I don't sin." They think that to be a sinner is to admit to being evil. This may have been the message that they got as a child. However, the Bible, God's Word, tells us that God loves the sinner. It states in the book of Luke, "There will be more joy in heaven over one sinner who repents, than over ninety-nine righteous persons who need no repentance." I like the following analogy that Jesus offers in Luke, "Those who are healthy do not need a physician, but the sick do. I did not come to call the righteous, but sinners." The latter statement makes me want to consider myself a sinner! He came for **ME!**

The reason why I desire to write on this topic is because this is the message that I would like to give to my father. I want to tell him that I forgive him. I want him to know the unconditional love that I have found in God the Father. I want him to know that God loves him. Yes, even while knowing all of the ways that he has offended me and our loved ones. I am afraid that he will never let Jesus live in his heart and experience His peace and joy, because he may feel unworthy. I want him to know that none of us are worthy of God's grace. It is a gift.

People are astounded that I am able to forgive my father and still love him. If the person is a parent, I ask them to imagine their child committing offenses or disappointing them. Then I ask, "Do you love them any less?" "Do you stop trying to be there for them to help steer them the way they ought to go?" The answer is always, "No."

This is my understanding of God the Father's love for me. I will tell you a tale, one that God gave to me. When I first began opening my heart to Jesus in my early twenties, I had a lot of shame. I knew I had made a lot of bad choices that God wasn't pleased with. Part of me questioned if I was loveable at all. However, God's Spirit that I welcomed to live inside of me started correcting these thoughts that I had.

God's Spirit described the truth to me in this tale…

There was a young girl about the age of seven. It was the start of a new day. It was a bright and sunny spring day and she was so looking forward to the day's special event. She donned her very favorite dress and all of the frilly accessories that go along with it. She sat patiently through the hair stretching ritual so that her hair would look "just right." When she was certain that she looked like a princess, she started off toward her destination. She rushed outside, only to fall into a muddy puddle left by last night's rain storm. She is covered in mud. It is as if her world has ended.

But wait a minute. Her father's big, strong hand is there. Unafraid of dirtying himself, he lifts her up into his arms and holds her. He tells her that everything is going to be all right and that she is still his beautiful princess. He takes her into the house and gets her all cleaned up. Eventually, she is as good as new. Her father then takes her by the hand and walks with her as they set off to their destination. As she walks outside, she has forgotten all about the fall in the mud. She is happy and looking forward to the day's event. After all, she feels loved and knows that she is not alone.

God created us. We are His. We are His precious children. He sent His son Jesus Christ to live amongst us to show us firsthand what Godly love is all about. I love the scenes described in the New Testament where Jesus spends time with the people that even the disciples' questioned. Jesus showed respect and acceptance to the adulterer, the dishonest, greedy, tax collectors, the prostitute, the leper, and the Samaritan woman, who had wed five times and lived unwed with a sixth man. Jesus even showed love to his long-time friend and disciple, Judas, who is also known as his betrayer.

Therefore, if Jesus was still living on earth and was in the area, he might choose to sit down and dine with my father. Who am I to say that God, who didn't spare His only Son, should sit with me instead? I cannot say that I am free of sin or of offending anyone. Jesus himself said, "Stop judging and you will not be judged. Stop condemning and you will not be condemned. In Luke 6:37, Jesus also says,

"Forgive, and you will be forgiven." I admit that I repeatedly fail in this area, so I have to ask God for the grace to live it. His grace is the reason why I can forgive my father—not in words alone. I am able to show respect and acceptance of my father, with his past and present faults; in my thoughts, words, and my actions toward him.

God fills me with compassion for my father, not knowing what it was like living in his shoes all the years before I was even born. God has given me His eyes—His heart, and so I am able to say, "Dad, I forgive you."

Words of Truth

"The Thief comes to steal, kill, and destroy. But I have come that they may have life, and have it to the fullest."

John 10:10

"You belong to God, children, for the One who is in you, is greater than
the One who is in the world."

1 John4:4

"Create in me in a clean heart O God, and renew a right spirit within me."

Psalm 51:12

"I will give them a new heart, and put a new spirit within them. I will remove the stony heart from their bodies, and replace it with a natural heart; thus they shall be my people, and I will be their God."

Ezekiel 11: 19

Author's Note

I have learned that forgiveness means LETTING GO. Letting go of the pain. Letting go of the anger. Letting go of the resentment. If you don't, you punish yourself, not the offender. I always felt that God wanted me to forgive my offenders, particularly my father, but it felt too hard. I thought that I couldn't do it. Actually, deep down inside, I didn't want to. If I forgave my father, then it would be as if I were letting him off the hook. This was especially hard because he never said he was sorry for hurting me. What helped me along this process was a story found in Luke 15:11-32, known as The Parable of the Lost Son. This is a very well-known story, but God showed me something new.

When the father saw his estranged son walking along the road toward home, he immediately proceeded to prepare for a feast in honor of his son's return. The father was wounded by his son's behavior, yet he didn't wait for the son to return to him and apologize. He had mercy on his son and forgave him. When his son reached his father, he did apologize, but the father didn't wait for that. "While he was still a long way off, the father caught sight of him and was filled with compassion. He ran to his son, embraced him, and kissed him" (Luke 15:20). God was trying to tell me to let go of all the ugliness. Just do it. Don't think about it. Don't pay heed to those old, familiar, dark feelings. Just LET GO.

Once I let go and made the decision to love my father from that moment forward without remembering his offenses, I saw a big difference in my marriage. Prior to forgiving my father, I had underlying feelings of hatred and fear of men, including my own husband. It wasn't easy for me to love my husband. I always saw his faults. I was always disappointed in the things he did or didn't do. But after I forgave my father, I behaved differently toward my husband. I found myself accepting my husband's faults—accepting him for who he was. I became a more loving wife. I was married for six years prior to this moment of change. I was able to see the difference that forgiving my father made in my life.

I hope that in sharing my experiences, others' lives may be affected for the better. That is my prayer for you.

About the Author

Holly and her husband Steve Otten founded Tin Man Ministries with the goal of promoting healing for survivors of sexual abuse and advocating for victims of the sex trafficking trade. In addition to writing *The Tin Man,* Holly has written five other faith-based books: *hand in Hand* (her life story), *Free to Dance, Made for Love, Brides for Christ* and *Holy Fire*. Holly has also created the *Healing Hearts* workshop, which is a Christ-centered 8-week workshop that takes survivors through the healing process.

Holly's 6 books are available on Amazon (print and ebook) and also can be purchased at www.tinmanministries.com/store.

To learn more, go to www.tinmanministries.com.

"They overcame him (devil) by the blood of the Lamb
and by the word of their testimony."
Revelation 12:11